Euro-American
Villa Soft Decoration

欧美别墅软装

深圳视界文化传播有限公司 编

中国林业出版社
China Forestry Publishing House

PREFACE 序言

I LOVE DESIGN AND IT ALWAYS MAKES ME HAVE A CREATIVE MIND

热爱设计，让我不断充满创造力

As a luxury residential interior designer, my job is to create homes that reflect the dreams and lives of the people who live in them. I create the stages in which their lives will play out. I don't believe in creating "a look" and repeating it over and over because I feel each of my clients is unique and deserves their own special look. My taste is my own and I reserved that for my personal home. I am not trying to make everyone live the same way I do but to create individual personal styles for each and every client I work with. To know and see how each person uses their personal space and the function of objects that surround them and to understand their psychology of design is always fascinating to me, meanwhile, it's the instinct of design.

I am sensitive to people's needs, to what they say they want and I try to take that information and go beyond it. I listen cautiously and approach design fearlessly by always finding bigger and better than what was asked for. One of the advantages of having a creative right brain thinking mind, is that I have more ideas than I can execute. So it is easy for me to continually create and recreate rooms. It is easy for me to add or delete and curate groupings of fabrics. I dream in color and 3D and my projects always floating around in my mind.

I am very focused on my clients' homes and my home always takes a back seat simply because my taste is also ever changing. I am attached to a few special pieces of art, and a few irreplaceable antiques that will always stay with me. These individual decorative items can represent my taste and preference, and as the main elements in the space, they are the most important factors to embody the characteristics. But overall there is no need to have strong attachment to material things for the simple reason that we are surround by beauty and spend so much time shopping for and creating little bubbles of perfecting as life passes by. Therefore, we should love design and enjoy the beauty brought by design. Beauty and interior design make us happy, which makes them more worthwhile.

作为一名豪华住宅的室内设计师，我的工作是创造出能够折射居住者梦想和生活的房子，我为他们的生活创造舞台。我不认为设计应该有某一种特定的"外观"，然后一遍又一遍的重复，因为我觉得每一个客户都是独一无二的，他们需要有自己的风格。我的品味是属于我自己的，因此我把它保留在自己的家里。我并不是想让每个人都像我一样生活，而是为每一个和我一起合作的客户创造她独有的个人风格。了解每个人如何利用他们的个人空间与他们周围各类物体的功能，以及理解他们关于设计的想法，这些非常吸引我，也是设计该有的本能。

我对人们的需求以及他们想要的东西很敏感，我试图去获取这些信息，并超越它们。我静心聆听，大胆设计，总能超出客户的期望。拥有一个创造性的正确思维方式的优势之一，就是我总会有更多的想法。所以对我来说，对空间进行不断地创造和再创造比较得心应手。增添或摒弃，以及对结构进行重组也是如此。我的梦境是彩色立体的，那些项目也总是萦绕在我的脑海里。

我非常重视客户的家，甚至超越了我对自己的家的关注，因为我的品味在不断变化。我喜欢一些独特的艺术作品，和一些能够一直伴随着我而又绝无仅有的古董。这些个性化的软装陈设，更能代表我的审美和喜好。软装作为空间的主要载体，是体现空间特色最为关键的一步。但总的来说，对物质的依恋并不需要太强烈，原因很简单，我们花费很多的时间去购买这些美妙的东西，而它们就像是泡沫，随着时间的流逝会慢慢消失。所以热爱设计，享受设计带来的美好！把美好的东西和室内设计变成让我们感到非常快乐的实质，更为值得！

Sandra Espinet
S.E. Design Services

CONTENTS 目录

006	Luxury and Elegance Accomplish a Kingly Demeanor 奢华典雅 缔造王者风范	*174*	A Beautiful Mix of Colors 斑斓色彩汇
022	Build an Artistic Castle 构筑艺术化的古堡	*186*	Inspiration of a Beautiful Home 家的灵感与美丽
040	Old Hollywood Glamour 古老的好莱坞魅力	*192*	An Elegant World in Colonial Style 殖民风格中的文雅世界
056	Wonderful Life in Golden Years 流金岁月 精彩纷呈	*202*	Exotic Influence 异域来风
066	Timeless, Classical and Exquisite Mansion 隽永经典 质感宅邸	*212*	A Luxury Home Filled with Happiness 幸福洋溢的奢华之家
074	Exotic Oriental Totem 异域风情的东方图腾	*224*	Extreme Design Interprets a Remarkable Temperament 极致设计 演绎非凡气质
083	Blueprint for Life 描绘生活蓝图	*234*	A Second Friendly Nature 第二个友好的自然
098	Enjoyment of the Ocean 赏尽大海风情	*242*	Customize the Home with Art 定制家的嫁衣
114	An Artistic Villa that Awakens Life 唤醒生活 艺术庄园	*252*	Fresh and Delightful Life 清新愉悦的惬意生活
126	Psychedelic Wonderland 迷幻仙境	*260*	A Young House that Releases Personality 释放个性的青春家
134	A Colorful and Warm House 缤纷多彩 温暖如初	*272*	Slow Life in a Rural Home 田园之家的慢生活
146	A Desirably Leisure House 向往之家 休闲温度	*280*	A Resort with Spectacular Sea View 辽阔海景 度假胜地
158	Retro Complements with Art 复古与艺术相生	*298*	Winter Sonata 冬季恋歌
166	Fragrance in Summer 芬芳夏日长	*308*	Dancing Party in a Youthful Manse 焕染青春的庄园舞会

Design Company / 设计公司：Haleh Design Inc
Designer / 设计师：Haleh Alemzadeh Niroo
Project Location / 项目地点：Maryland, USA
Area / 面积：1858m²
Photographers / 摄影师：Gordon Beall, Greg Powers

Luxury and Elegance Accomplish a Kingly Demeanor

奢华典雅 缔造王者风范

The luxurious French style home decorations present a strong aristocratic artistic and cultural atmosphere. Baroque and Rococo styles have three or four hundred years of history, they reflect unique charm of the times, and are known as the representatives of luxury romanticism, which are the most classic styles. The elegance of Baroque style reduces the extravagance of the golden color, while the softness and exquisiteness of Rococo style make it more sophisticated and gorgeous, this is the perfect interpretation of French luxury.

　　法式奢华家居富含浓郁的贵族宫廷色彩，带有艺术与文化气息。已有三四百年历史的巴洛克和洛可可风格凝练出独特的时代韵味，被誉为奢华浪漫主义的代表，成为经典中的经典。巴洛克风格的优雅大气使金色摒弃了土豪气质，洛可可的柔美精致让金色更为精巧华贵，是诠释法式奢华的完美结合

The essence of French Baroque style is the graceful and flowing line, and the main features are the gorgeous sculpture, exquisite mosaic and gold-plated approach, together they create a classic, luxurious and romantic style. In the entrance hall, the orderly symmetry conveys a sense of an integral simplicity, four elements are depicted around the 32-feet-high dome by murals, implying the earth, the wind, the fire and the water. The curves of the staircase combine solid mahogany handrails and hand-forged iron railings with 18k gold leaves, columns and cornices are also covered with 18k gold leaves. At the same time, the wallboards which are inlaid within the silk add warm feeling to the space, the complex carvings show the infinite fantasy of art design, and the classical decorative patterns preserve the Baroque style's elegant soul very well.

优雅的流动线条是法国巴洛克风格的精华，以华丽的雕刻、精巧的镶嵌和镀金手法为主要特征，打造出经典的奢华浪漫主义风格。门厅的有序对称传达了一种整体的简单感，围绕着9.8米高的圆顶壁画描绘了四种元素，寓意大地、风、火和水。楼梯的曲线结合了坚实的红木扶手和手工锻造的铁栏杆，带有18k金叶，柱状首府和檐口在18k金叶中。同时，镶在丝绸上的镶嵌壁板增加了温暖，繁复的雕刻展现了对艺术设计的无穷幻想，古典装饰图案完好地保留了巴洛克的优雅灵魂。

Gold is definitely used in this extravagant space. A large area of gold in different shades interprets a gorgeous atmosphere of the court style, whilst a few wood color makes the study space not too dazzling. But the large leaf crystal chandelier makes the place more elegant and luxurious.

纸醉金迷的奢华无法摆脱金色的烙印。不同深度的大面积金色共同演绎华丽大气的宫廷风格，少量的木色使得书房空间不会过度耀眼。加入大型叶子水晶吊灯，更富有典雅与奢华。

The noble and elegant flavor of a European style villa is hidden in each space, which brings us positive energy and shows our advocacy of gorgeousness. Fine carvings, delicate white and blue colors highlight the luxury and elegance of a kingly demeanor. You can see the blue sky, white clouds, blooming grasses and trees, and you will live a glittering life in this colorful world, and such beauty will become an eternal history.

欧式别墅的高贵典雅之气蕴藏于每个空间之中，带给我们积极的正能量和华丽的尊崇。精致的雕刻、细腻的白与蓝，彰显出奢华优雅的王者风范。蓝天白云，草木相生，让这样的五彩斑斓，伴你走入金光闪闪的年华，完成历史的永恒。

Build an Artistic Castle
构筑艺术化的古堡

Designer / 设计师：Vivian Reiss
Project Location / 项目地点：Toronto, Ontario, Canada
Area / 面积：697m²
Photographer / 摄影师：Derek Shapton

The designer created this home to showcase her love of family and creations. She is a lifestyle expert and loves to create artwork, food, and gardening. The residence was once the home of the prestigious Baldwin family in the 1900's and she completely gutted the home while still keeping the integrity and grandeur. Her love of color and opulence created a beautiful maximalist home.

设计师创造出这个家以展示她对家庭和创作的喜爱。她是创造生活方式的专家，喜欢创作艺术品、食物和园艺。这个住宅曾经是名门望族鲍德温家族 1900 年代的家，设计师在保持房子原有的完整性和庄严性的同时，彻底整修了它。以设计师对色彩的热爱和专业的知识来创造一个丰富美丽的极多主义之家。

This is her personal home and she wanted to fill it with the collections from her travels from around the world - beds from Bali, statues salvaged from old newspaper buildings, and windows from a cathedral. She loves colors and salvaging materials from other buildings, through her love of this she inadvertently created a maximalist home.

这是设计师的私人住宅，她想把她从世界各地旅行中收集到的藏品放入家中——从巴厘岛买的床、从旧报社建筑中回收的雕像以及教堂的窗户等。设计师喜欢色彩，也喜欢从其他建筑中回收材料，这些东西无意间为她创造了一个极多主义的家。

The diversified and rich colors accord with the designer's hobby and taste. Through the purplish blue background, we can feel the noble, ocean-like profound temperament, as well as the gentle and elegant gesture. People always like blue, because its quietness and steadiness make us feel peaceful and clam, and when it matches with bright white, they achieve a simple style, yet reveal silent power. Meanwhile, the golden color reasonably balances the collision between cold and warm colors.

室内色彩多元且丰富,符合设计师的爱好与格调。大面积藏蓝色背景基调中,有着高贵的血统,海洋般的深邃气质,绅士般的优雅姿态。自古以来,便是被世人宠爱的色彩。静谧而沉稳的气质蓝,能够令人凝神静气,从容心安。与亮白色搭配简约而具有沉静的力量。同时,金色的融入,将冷暖色系的调性撞击达到空间的理性平衡。

She loves bright colors and chose turquoise for her master bathroom, baby pink for her daughter's room, and light green in her living room. It is important to Vivian to have a space to channel her creativity. Her garden, painting studios, and kitchen are some of her favorite parts of the home. She wanted to make her bedroom and her now grown children's old bedrooms bright and colorful and with special additions, like the candy molds that she used to adorn the ceiling and walls in her daughter's old bedroom. She turns a home into an artistic castle, and this is the charm of design!

设计师喜欢鲜艳的颜色,于是将主浴室选择了绿松石色,女儿房是粉红色的,客厅里是淡绿色的。对她来说,有一个空间来发挥她的创造力是很重要的。花园、绘画工作室和厨房都是她最喜欢的部分。她想把卧室和她现在已经成年的孩子们的旧卧室变成明亮多彩的,因此选了一些特别的东西来装饰房间,比如她用糖果模具来装饰她女儿旧卧室的天花板和墙壁。将家打造成一座艺术化的古堡,这就是设计的魅力!

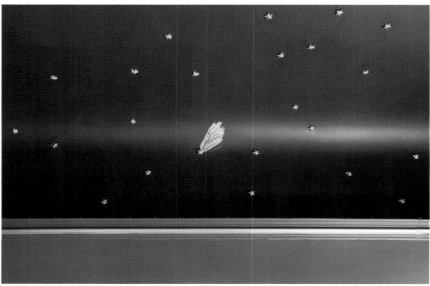

Old Hollywood Glamour
古老的好莱坞魅力

Project Name / 项目名称：Malvern Residence
Design Company / 设计公司：Massimo Interiors
Designer / 设计师：Massimo Speroni
Project Location / 项目地点：Malvern, Melbourne, Victoria, Australia
Area / 面积：533m²
Photographer / 摄影师：Stu Morley

The designer made a space that presented the "old Hollywood Glamour", which injected with bursts of color and audaciously bold with a touch of Renaissance! Every piece of furniture is very valuable: the Center Table at the entrance is an early 19th Century Italian inlaid table; the gold frame in the living room is actually an original 19th century French Louis XVI style mirror; the twin swan-tub chairs in the lounge have a modern take on these period Napoleon III empire style chairs…He was aiming to create a space that reflected the clients' personality maintaining an elegant and sophisticated atmosphere and gave them an artistic temperament.

设计师围绕着创造古老的好莱坞魅力为设计主题，为这个空间注入大量的色彩，打造文艺复兴时期的既视感。每一件家具，设计师都如数家珍——入口处中间的桌子是19世纪早期的意大利镶嵌式餐桌，客厅的装饰金架实际上是来自19世纪法国路易十六的镜子，休息室的天鹅式椅子灵感来自拿破仑三世帝国风格的椅子……设计师的目标不仅是创造一个能反映客户个性的空间，同时能保持优雅而精致的氛围，给予空间一种艺术的质感。

The wall mirror was one of his key design solutions for the lounge. The challenge was to infuse light and brightness into the connecting and dark formal dining and lounge room. The introduction of a panelled wall-mirror, metallic wallpaper, and white surfaces helped reflecting and amplifying the light coming into the room throughout the day. The gold frame is actually an original 19th century French Louis XVI style mirror of excellent quality. It retains the original gilt finish and all the original mirror plates. The frame is decorated with bead-and-leaf motifs, the crest with out-sized laurel wreath, a knotted bow and crossed quiver and torch motifs. The wall mirror needed a center piece, a focal point above the sofa; hence the introduction of this stunning piece. It creates a layering affect enhancing the height of the ceiling yet maintaining a light classical atmosphere in the room. It is balanced either side by the 2 Murano glass wall sconces reinforcing the symmetric layout in the room.

墙镜是设计师设计这个房间的关键策略之一。设计的挑战在于如何将光线和亮度引入到稍暗的正式餐厅和休息室。一个镶板的壁面镜子、金属壁纸和白色外表面的引入，使白天进入房间的光线大大增多。墙镜上的装饰金架实际上是来自19世纪法国路易十六的镜子，它保留了原有的镀金和所有的原装镜板，镜框上装饰有珠状的叶子图案，顶部有大号的月桂花冠、一个蝴蝶结、交叉的箭筒和火炬图案。墙镜不仅是一件令人惊叹的作品，更是在沙发上方创造一个视觉中心点，创造了一种分层效应以提高天花板的高度，同时保持了室内的古典气氛。两个平衡的 Murano 玻璃墙壁突式烛台则加强了房间的对称布局。

The wife's brief for this generous size master bedroom was to achieve a sexy "Boudoir" bedroom. As this room needed to flow with the design and style applied to the rest of the house, the designer came up with a modern and eclectic interpretation of a classical "Boudoir" bedroom still keeping the old Hollywood glamour atmosphere. The dark purple headboard and patterned carpet respond to the color of the ceiling, together they create a noble and graceful feeling.

女主人想在这个宽大的主卧室里安插一个迷人"闺房",鉴于这个房间的设计和风格需要照应房子里其他空间的风格,设计师想出了一个现代的折衷的诠释手法,成功在古典的"闺房"卧室里保留了古老的好莱坞魅力气氛。深紫色床头与印花地毯及整个天花的颜色相互呼应,共同营造了高贵典雅的空间氛围。

Parents fear that their kids will quickly grow out of a specific themed bedroom, with the impact of further expenses to redecorate their bedrooms after a year or two. So rather than committing to a particular theme, the designer selected appropriate furniture according to the children's usage and personality. For one of the rooms, he sourced a graphic yet subtle wallpaper for the ceiling and hung it on a 45 degree angle for a more dynamic effect. It creates a visual interest and it changes its direction depending on where you stand in the room. This makes the children very satisfied.

父母担忧卧室主题更新的速度跟不上孩子们成长的速度，那样势必会在重新装修设计儿童卧室上带来更多的开销。因此设计师在设计儿童卧室时并没有选择一个特定的主题，而是根据孩子的使用模式，结合孩子的性格，选择合适的家具。如在其中一个房间，天花板上贴了图形化的、精细的墙纸，把它挂在45度角上，能获得更动态的效果。它会产生视觉上的效果，会根据你在房间里的位置而改变它的方向，这让孩子们十分满意。

Wonderful Life in Golden Years
流金岁月 精彩纷呈

Project Name / 项目名称：Jerusalem Penthouse
Design Company / 设计公司：Geoffrey Bradfield, Inc.
Designer / 设计师：Geoffrey Bradfield
Main Materials / 主要材料：Woods, Fresco Marble, Carpet, etc.

The golden strings are playing a melodious life music as time passes by. Who opened the history? Who depicted the trails of time? We are listening to the blessing bell, and enjoying the sunshine which is just like having a holy baptism. The integrated colorful shadows transfer the immortal soul, and the golden years belong to us.

悠扬的琴声穿透岁月，命运献上金色琴弦，时间演绎这一切。是谁揭开历史的帷幕，是谁刻画了时代的金色年轮，我们聆听着祝福的钟声，沐浴着阳光的神圣洗礼，交织地斑斓影迹，传递不朽的灵魂，流转光芒的金色年华，正是属于我们的黄金时代。

The brilliance of gold attracts all the attention. As a symbol of wealth, gold is applied to the luxurious decoration and embodied in colors, so that we can see various kinds of golden adornments, such as collections, hanging pictures and others, all are shining brightly and being loved by people.

黄金的耀眼光泽，吸引全部的视线。作为富贵的象征，运用于奢华的家居设计，体现在色彩上，各式各样的金色装饰物件应运而生。如收藏品、挂画等，熠熠生辉，无法阻挡人们对它的喜爱。

A mix of red and beige creates an exquisite, romantic and fashionable bedroom. All sorts of red items introduce an intense exotic feeling, whilst the integration with comfortable beige makes a warm and refined picture.

红色与米色相互糅合,为你打造精致浪漫的时尚卧室空间。各种红色的饰品置于其间,带来异域风情的热烈,与舒心的米色构成一幅温馨雅致的画卷。

Project Name / 项目名称：Morningside Legacy
Design Company / 设计公司：Christopher Kennedy
Designer / 设计师：Christopher Kennedy
Project Location / 项目地点：Columbia, USA

Timeless, Classical and Exquisite Mansion
隽永经典 质感宅邸

The herringbone wood dropped ceiling increases the height of the space, and the wood color not only makes the space look spacious and transparent, but echoes with the carpet. The decent sofas have a light color tone and the furniture have simple lines, together they easily create a leisure and free living room. The owner wanted to be close with the outdoor scenery, so a lot of glass materials have been used in one side of the walls, providing unobstructed views for people. Without too much decoration, everything is as natural as the outside scenery.

人字形条纹木的吊顶，拉伸了空间高度，木色也使空间显得宽敞通透，同时与地面地毯颜色相呼应。素色大方的沙发与简洁家具线条，轻松打造出休闲自在的客厅空间。屋主想要与户外美景无限贴近，所以在客厅一面用大量的玻璃建材，使得居者可以无阻碍观赏户外风景。没有过多的装饰，一切都与外面的景色一般自然。

The designer chose Chinese style products to adorn the space. The Chinese style kitchen shelf becomes a basis for the decoration, on which there are elegant and beautiful ancient potteries that have Chinese classic patterns, bringing a historical and cultural connotation. The large crescent-shaped cyan sofa breaks the convention, gives the space a new interesting look and provides people an impressive visual experience.

设计师挑选极具中式风韵的单品来装饰空间，中式橱架成为软装基础构架，清丽素淡的古典陶罐穿插其中，再辅以中式经典纹样，带来一种历史的内涵与文化的积淀。占据大半视野的月牙形青色沙发，打破常规的轮廓造型赋予空间新奇的观感和趣味，带来眼前一亮的视觉感受。

There are not many furniture in this concise and modern bedroom, and a large area of free space in the middle makes people feel comfortable. The overall color scheme is very simple, and the background color of pure white sets a soft and quiet tone. Light cyan is used for the lazy sofa and bed stool which have graceful curves, while the charming light purple is applied to the large size bed. The combination of the three elegant colors expresses a refreshing home atmosphere.

在这个简约现代的卧室中,并无许多家具填充摆放,中间留出大块空间,宽大而自由的空间让人心生惬意与自在。整体用色十分简洁,以纯白作为背景色彩,为空间定下柔和宁静的基调。淡青色则用于曲线优雅柔美的懒人沙发及床尾凳,大床使用迷人的浅紫色,这三种淡雅的色彩,组合在一起,烘托出清爽治愈的家居氛围。

Exotic Oriental Totem
异域风情的东方图腾

Design Company / 设计公司：Freimann architectural bureau
Designer / 设计师：Olga Freimann
Project Location / 项目地点：Moscow, Russia
Area / 面积：500m²
Photographer / 摄影师：Dmitriy Livshits

The owner of the apartment wanted to implement the colonial style in the interior, to combine European and Eastern elements in decoration, especially some motives of Moroccan style, which are a little bit exaggerated, but distinctive. Another task for an architect was to install the big collection of paintings and antiques, that's why the context must be appropriate for them. To fulfill this task, the architect decided to use the aging columns and mantelpiece of fireplace in living room. There are a lot of handmade pieces in the apartment: paintings on the walls, specially designed furniture, fireplaces, mosaics and wallpaper. Besides, the handmade prints using authentic matrices from the 18th century show a strong classical flavor.

主人想将室内装饰成殖民风格，并将欧式元素和东方元素结合在一起，尤其是摩洛哥风格的一些图案，虽然夸饰但也独具特色。建筑师的一个重大任务是将大量的绘画和古董摆放在室内适当的位置以符合屋主的生活情境，为了完成这一任务，建筑师决定将老化的石柱和壁炉安放在客厅。公寓里也有很多手工制作的作品，比如说墙上的画、专门设计的家具、壁炉、马赛克和墙纸，同时18世纪原汁原味的母岩手工制作的印花也被完整地保存下来，装饰着厚重的古典韵味。

Another idea of this interior – the eastern motives in classical design. They are represented by a small table with dragon image at the hall, painted plafond at the office room, door's mantels. Besides there is a lot of gold in this interior which symbolizes the sun. But the gold here is muffled, for this reason it was also agented. Beside the gold, mirrors with bronze create a warm atmosphere.

另一个室内装饰的理念即东方图案在古典设计中的运用，大厅里的一张画着龙图腾的小桌子、办公室的装饰天花板、门框等地方都有所体现。除此之外，室内还有很多象征着太阳的金色，但由于大量金色会让空间变得沉闷，于是设计师将一些金色用其他原料替代，而金色周围，青铜制成的镜子则营造了温暖的氛围。

Project Name / 项目名称：Dallas Project
Design Company / 设计公司：Kirsten Kelli LLC
Designers / 设计师：Kristen, Kelli
Project Location / 项目地点：Dallas, USA

Blueprint for Life
描绘生活蓝图

This is a house with museum-quality art - including pieces by Matisse, Picasso, Koons and Magritte co-exist with two cats and five dogs. Kirsten and Kelli balanced the grandeur of a Dallas house with an unabashed passion for color, used classical touches and bold colors to strike a balance between formal elegance and whimsy. The work of the late David Hicks inspired the geometry of the gallery floor. The owner has high demands on the house, it must be grand enough to accommodate a hundred dressed-up guests for cocktails on a Friday night, kid-friendly enough to host a gaggle of eleven-year-olds at a Saturday afternoon pool party, and cozy enough to serve as the spot where the Fords can kick back on a Sunday to watch a Cowboys game.

这是一间有着博物馆级艺术品的住宅——包括 Matisse、Picasso、Koons 和 Magritte 的作品，还有两只猫和五只狗。Kirsten 和 Kelli 毫不掩饰地用色彩衬托了达拉斯住宅的宏伟，用经典的基调和大胆的色彩让正式的优雅和奇思妙想之间达到平衡，用已故的 David Hicks 的作品启发了画廊的几何图形地毯。屋主对住宅有着高要求的标准，住宅空间必须宽敞大气以满足容纳一百多人用餐的周五鸡尾酒晚宴，必须有孩子喜欢的氛围从而在周六下午举行儿童泳池派对，必须足够舒适以作为周日看牛仔游戏的场地。

This artistic house has lots of rest and leisure spaces, such as the fresh family room, in which the light blue color creates a different vision. The light blue clapboards have golden edges, which are echoed with the golden leather tea table that has rivets. The color scheme of the family room paid great attention to the visual effect, making people feel enjoyable when entering the room.

这座艺术住宅有许多可以供给家人休息休闲的空间，譬如充满清爽气息的家庭娱乐室，以浅蓝色为肤打造不一样的视觉空间。浅蓝色的护墙板以金色镶边，金色皮质的茶几与之相呼应，上面还精心镶嵌了铆钉。家庭休闲室的色彩搭配十分讲究视觉上的美感，让人走进房间便不自觉得身心愉悦。

Enjoyment of the Ocean
赏尽大海风情

Design Company / 设计公司：S.E. Design Services
Designer / 设计师：Sandra Espinet
Project Location / 项目地点：Cabo San Lucas, Mexico
Area / 面积：929m²
Photographer / 摄影师：Hector Velasco Fazio

This total renovation project in Cabo San Lucas, Mexico was designed to maximize the beautiful ocean front property and to create a personal modern space filled with beautiful art to create the perfect family vacation home. The original home was dark and dated. We demolished most of the interiors and all windows were made larger and we added in huge slidding glass doors to each ocean facing room to create lots of light so that the ocean could become the star. Mother nature is the best designer of all, so we need to live in harmony with nature, and use it to design beautiful scenery. From completion to end it took about 16 months to finish this job. Our firm was involved from start to finish and we did all details from interior architecture, specification, lighting, furniture all the way to accessories like dishes and hangers.

这是一个翻修项目，位于墨西哥的卡波圣卢卡斯。设计目的是最大限度地利用其所处的美丽沿海位置，并创造出一个拥有美丽艺术与个性化的现代完美家庭度假屋。原来的房子很暗且年份较久，我们拆掉了大部分的室内构架，扩大了所有窗户的面积，并在每个面朝大海的房间里添加了巨大的滑动玻璃门，以引入大量的光线，将海洋景观变为空间的常景。我们都说大自然是最好的设计师，我们选择与自然和谐相处的同时，再充分利用大自然来构筑美景。整个项目从开始到结束，团队花了大约 16 个月时间来完成。从室内建筑、尺寸规格、照明设计、家具选择等各方面都做了详细的前期准备，整个项目设计公司从头到尾都积极参与。

The home was first approached from the architectural point of view and all finishes, tiles, stone floors, plumbing fixtures, carpentry designs and materials were selected. Light colors, creams, sands and teals were used because they reflected the outside ocean and beach tines. We used exotic materials such as eel skins, mother of pearl along side modern materials such as laquered woods and glass, they are either special or fashionable.

设计师首先从建筑的角度入手，选定了所有的瓷砖、石材地板、管道设备、木工设计等材料。同时，在色彩方面定义为浅色，通过漆面、沙子等方式，与外面的海洋与海滩相呼应。另外在选材上使用鳗鱼皮、珍珠母这样异域风情的材料与漆木、玻璃等现代材料相结合，使得个性与时尚在这一方天地间自由切换。

The client wanted the inside to be harmonious with the outside natural exteriors. After that the interior furnishings, fabrics and art were selected and were treated with a bit of a modern twist. The client did not want stock, standard modern furniture that looked like an Italian showroom. She wanted warm, personal and very user friendly furniture that everyone could use and enjoy even in bathing suits. Luxury lines such as Donghia Furniture, and Circa Furniture were used to create unique custom pieces and create the incredible open spaces that make up this home. The mother of pearl cabinets in the master bedroom were made to order from manufacturers in the Philippines and all of the custom carpentry such as the 3D sculptural living room wall, were made locally in Mexico.

客户要求室内装饰设计与外部自然环境和谐一致，于是室内陈设、艺术品及面料选材都被挑出来，并做了相应的现代化修饰。客户希望自己的房子是独一无二的，不想要常见的、标准化的现代家具，让房子看起来像一个意大利的陈列室。她希望家具是有温度的、个性的和舒适的，并且每个人都可以使用，甚至可以穿着泳衣漫步其中。因此选择了独特的 Donghia 和 Circa 这样奢侈家具品牌的定制品，创造出精美绝伦的开放空间。主卧室里带珍珠的橱柜是在菲律宾的制造商那里订购的，所有定制木工，如 3D 雕刻的客厅墙壁，都是在墨西哥当地制作的。

A poetic and fine place is the best description for the bedroom. The tone is simple, the bed products are fresh, and the embellishment of flower plush not only vitalizes the white space, but also embodies a distinct flavor.

In the dazzling guest room, the elegant light peacock blue geometric carpet is a combination of fashion and romance, which can absolutely catch anyone's attention. The bright wooden suspended ceiling, clean and comfortable bedding and pillows give us a strong retro feeling. The whole bedroom presents a striking effect by fresh colors, meanwhile, reveals a noble, refined and romantic taste.

用诗意栖居的美好意境来形容这间卧室空间真是再合适不过了。简约的格调，清爽的床品，花朵毛绒的点缀与亮白色结合，不仅让白色空间灵动起来，还演绎出一种别样风情。

在客房内，优雅的浅孔雀蓝几何地毯是时尚与浪漫的结合，绝对能吸引任何人的眼球。搭配光亮的木质吊顶、干净舒适的床品及印花抱枕，给人一种扑面而来的浓郁复古气息。整个卧室设计既有清新的色彩冲击，又不失高贵雅致的浪漫。

Design Company / 设计公司：David Kensington Interior Architecture & Design
Designer / 设计师：David Kensington
Project Location / 项目地点：Atherton, California (Silicon Valley), USA
Area / 面积：2322m²
Photographer / 摄影师：Cesar Rubio

An Artistic Villa that Awakens Life
唤醒生活 艺术庄园

The Atherton Villa, a 25,000 square foot new construction with 7 bedrooms, 7 full baths and three powder rooms is set on a 1.5 acre lot with mature oak and redwood trees and landscaped with fruit trees and boxwood. The residence is modeled after one of the most famous houses in Italy–Villa Almerico Capra Valmarana, which was one of the most inspirational architectural prototypes throughout the world for 500 years and is conserved as part of the World Heritage Site.

这个新建的2322平方米的阿瑟顿庄园有7间卧室、7套完整的浴池和3间化妆室，位于占地1.5英亩的土地上，有参天茂盛的橡树和红木树，还有果树和黄杨木，葱葱郁郁的树木赋予了庄园更舒适的环境和赏心悦目的风景。庄园的设计灵感来源于意大利最著名的住宅之一——Villa Almerico Capra Valmarana，500年来它被人们称赞为世界上最具启发性的建筑原型之一，并被列入世界文化遗产名录。

The Atherton Villa is a masterpiece of quality, the architecture and construction are exemplary in every detail. The grand residence simultaneously integrates traditional materials and the finest craftsmanship sourced from throughout Europe including 18th century French palace interiors; hand carved wood paneling, hand carved plaster ceilings, custom hardware by David Kensington in collaboration with P.E. Guerin, New York throughout, adapted into a modern, well-organized building designed for the 21st Century that accommodates the needs of the contemporary family living within it.

阿瑟顿庄园是一个鬼斧神工之作，建筑和建造在每个细节上的表现都堪称典范。这个大住宅同时融合了传统材料和来自整个欧洲包括18世纪法国宫殿室内装饰的最好的工艺。手工雕刻的木镶板和石膏天花板、David Kensington和纽约P.E. Guerin合作定制的家具等，使得它最终被改造成一个现代的、结构优良的为21世纪而生的建筑，符合现代家庭生活的需要。

Solid Rosa Verona mantle in living room and columns in colonnade, sourced and hand carved in Italy. The ceiling is set with white squares which play a smooth rhythm. A light yellow carpet was put on the rift and quarter sawn, hand scraped white oak wood floors, creating a warm and harmonious feeling. Custom made 24k gold plated neoclassical light fixtures and sconces throughout, designed by David Kensington and made by P.E. Guerin NY add artistic ambience into the living room.

客厅结实的 Rosa Verona 地幔和廊柱，源于意大利手工雕刻，白色方格天花也有着整齐的韵律。纹理直略交错的手刮橡木地板铺上浅黄色的地毯，温暖又和谐。由 David Kensington 设计、纽约 P.E. Guerin 制作的 24k 镀金新古典灯具和壁突式烛台，更加增添了客厅的文艺气息。

Large logia along back of house accessible through French doors from living, dining and media rooms. Driveway and courtyard with solid travertine cobblestones set on concrete slab spaced to allow for optimal drainage, parking courtyard covered in decomposed granite. Custom designed bronze and glass front entry door, solid mahogany doors and windows with copper screens and pocketing cooper screen mahogany doors that retract into extra thick walls throughout, looking beautiful and gorgeous.

  从法式大门到客厅、餐厅和媒体室都可以看到屋后的自然景观。车道和庭院有铺着坚固的石灰和鹅卵石的混凝土板以提供最佳的排水功能，停车场的庭院也铺有风化花岗岩。定制的青铜和玻璃入口门前，坚实的桃花心门带有铜屏风的窗户，布艺库珀屏风的红木门嵌入厚实的墙壁中，更显美观和大气。

Psychedelic Wonderland
迷幻仙境

Project Name / 项目名称：Avenue Niel Apartment
Designer / 设计师：Géraldine Prieur
Project Location / 项目地点：Paris, France
Photographer / 摄影师：ALEXIS NARODETZKY Photographe

"Red is the color of passion, life, luxury, sensuality and desire." Géraldine Prieur, interior and furniture designer, based in Paris, France, said this. The apartment is located in the heart of a beautiful Paris neighborhood. The psychedelic wonderland serves as a home to her and her family. The door to the apartment opens to an eruption of brilliant colors. Moving from one room to another, one discovers Géraldine's the colorful and non-conformist universe. An unrivalled colorist, she has opted for a vibrant palette of color, energizing and full of pep, composed to exploit light by applying the additive color synthesis principle which superimposes the three primary colours: red, green and blue. Géraldine deifies all colors but red especially captures her favor because of its power, its emotions and the energy it unleashes.

"红色是激情、生命、奢侈、感官和欲望的颜色。"来自法国巴黎的室内家具设计师 Géraldine Prieur 这样说。这间公寓位于美丽的巴黎社区中心，"迷幻仙境"是她和家人的家。通往公寓的门开启了绚丽的色彩世界，从一个房间走到另一个房间，你会发现一个 Géraldine 的色彩斑斓、另辟蹊径的宇宙。作为一名无与伦比的配色大师，她选择了充满生机的调色板，富有活力和生机，通过运用叠加红绿蓝三种基本颜色的色彩合成原理利用光线。Géraldine 几乎使用了所有的颜色，但红色尤其受她的青睐，因为它所释放的力量、情感和能量，更具感染力。

The entrance hall provides a colorful welcome to the apartment. The original stained glass windows have been restored to their floral decorations in tones of amber and Venetian yellow. The bright fuchsia walls are offset with a green wall-to-wall carpet which gives the room a cool, calm appearance. The living room is painted in a striking aquamarine blue. It is filled with a combination of contemporary and classic furnishings and artworks. The Up To You console in the family room was inspired by a 1930s ring given to Géraldine by her grandmother, which is very meaningful and valuable.

入口大厅为公寓提供了丰富多彩的欢迎之礼。最初的彩色玻璃窗已经被换成琥珀色和威尼斯黄色的花朵装饰。明亮的紫红色墙壁与绿色的墙纸相互衬托，带给人一种凉爽、平静的观感。客厅被漆成醒目的宝石蓝色，摆放了当代经典的家具和艺术品，客厅里的控制台灵感来自于 20 世纪 30 年代 Géraldine 的祖母给她的戒指，具有非常深厚的纪念意义和情感。

The pink and green tones of the dining room recall the sweet tastes of our childhoods. It evokes romanticism and coolness. The kitchen is a living space where the family likes to gather and talk. The herringbone parquet has been painted black whilst the moulding, cornices and fireplace have simply been retained in their original state. An atmosphere of harmony and serenity reigns here. The Alizarin red of the master bedroom evokes passion, and symbolises love, emotions and temptation. Finding it rich, powerful and energising, Géraldine and her husband have always favored this bold hue for their most intimate of spaces.

餐厅的粉色和绿色色调让我们回想起童年时代的甜蜜滋味，它能唤起浪漫和冷静。厨房是一家人喜欢聚在一起聊天的地方，人字形的镶板被漆成黑色，而檐板和壁炉则被简单地保留了原态，营造了一种和谐和宁静的气氛。主卧室的茜素红唤起了激情，象征着爱、情感和欲望。感受到红色的丰富、强大和充满活力后，Géraldine 和她的丈夫总是喜欢在他们最亲密的空间中使用这种大胆的色调。

Design Company / 设计公司：David Dalton Inc.
Designer / 设计师：David Dalton
Project Location / 项目地点：Beverly Hills, California, USA
Area / 面积：1022m²
Photographer / 摄影师：Kelly Marshall

A Colorful and Warm House

缤纷多彩 温暖如初

Upon entering the villa, the furnishings that have various shades welcome you with a warming atmosphere. The designer created this house with exquisite design techniques and keen observation, aiming to redemonstrate the meaning of home for people, namely, it's not only a resting space for body, but also a shelter for spirit. Therefore, warmth is the most important element. No matter it's the fresh green, bright yellow, passionate red, or fantasy yellow green, all add the warmth of summer sunshine, and you'll become happy and delighted when you walk inside.

走进别墅室内，便能感知扑面而来的温暖气息，这是由不同暖色调的软装配饰呈现出来的。设计师用细腻的设计手法和敏锐的观察力为屋主打造这座房子，意在还原家对于人的意义，不仅是身体休息的栖居之地，也是精神上的避风港湾。因此，温暖便是它最重要的代言词。无论是浅绿的清新、亮黄色的绚丽、正红色的热情，还是黄绿色的自然虚幻等，都为家增添了一缕如夏日阳光般的温暖，当你走进室内，心情不自觉也变得明朗喜悦和开阔起来。

Here, all the European style furniture are wearing colorful clothes. Sofas and chairs are decorated with different colors, such as bright yellow, light green, yellow green, pinkish purple and bright red, besides, the curtains in each room show distinctive charm. Designer's subtle and smooth usage of colors not only breaks through the original features of European style furniture, but adds glory into the spaces and brings warmth.

在这里，欧式风格的家具在设计师的打磨下，都穿上了绚丽多彩的服饰。光是沙发和椅子就有明黄、浅绿、黄绿、粉紫和正红色等多种配色，每个房间的窗帘也各有形态。设计师对色彩敏感又细腻的运用，不仅突破欧式家具原有的风格，也让空间色彩散发出自己的光芒，为家增添回归初心的温暖。

Except several leisure rooms, the bedrooms occupy a large area. Compared to the colorful and lovely kid's room, the main bedroom is gorgeous and dignified with a more cautious and unified color scheme. The white ceiling is sculptured by plasters, looking concise but not simple. The grey walls are well united with the carpet, however, the orange yellow bedding breaks such harmony and introduces lively and vivid feelings to the mature and elegant room.

除去多间休闲室，几间卧室也占据了不少面积。相比儿童房的靓丽活泼，童真可爱，主卧室则给人以大气端庄之感，在色彩运用上也显得更加谨慎而统一。白色的天花板加以石膏雕刻，简约而不简单，灰色的墙面和地毯连成一片，但橙黄色的床品加入其中，打破这统一之感，使主卧室成熟中带着俏皮，优雅中又见活泼。

A Desirably Leisure House
向往之家 休闲温度

Project Name / 项目名称：Far Hills
Design Company / 设计公司：Diane Durocher Interiors
Interior Designer / 室内设计师：Diane Durocher
Builder / 建筑师：Polo Master Builder
Photographer / 摄影师：Peter Rymwood

The design company customized the interior style for the owner, as a result, a scheme that combined a comfortable, leisure and retro feel with a fine and exquisite taste won the praise of the owner. Three floors were orderly planned, the first and second floors are mainly for public areas and rest areas, whilst the underground is a leisure space for the owner to relax the spirit. The tough wall panels, ceilings and gypsum lines are very clear. The country style antique chandeliers, European style leisure furniture, carpets, green plants and flowers soften the space, so that this home is full of warmth and comfort.

设计公司为屋主量身定制了这座别墅的室内风格，以整体的舒适休闲复古之感和局部的精致细腻赢得了屋主的赞赏。三层空间有序规划着格局，一二楼以主要的公共领域和休息领域为主，地下空间则是主人精神的休闲之地。护墙板、天花和石膏线的硬朗，使空间线条明朗清晰。乡村风格的复古吊灯，欧式略带休闲的家具，地毯绿植和插花的装饰，又起到柔化空间、增加温度的作用，使家充满了温情和舒适之感。

After you enter the room and pass the corridor, you will see the living room which shows a gorgeous and formal sense. The raised double height design increases the magnificence of the living room without causing aggressive or unhappy feelings, but it creates a comfortable and elegant atmosphere through the contrast of white walls, light-colored sofas and carpets. Walls are decorated with lamps which are echoing with the ceiling, adding a court-like temperament.

　　进入室内，穿过廊道，便能感受到客厅颇有仪式感的大气和正式。挑高双层的设计向上可增加客厅的气势，但这气势并不显得咄咄逼人，让人不悦，而是在白色墙面和浅色沙发、地毯的映衬下显示出一种舒适和典雅的观感。四周墙面上环绕装饰的壁灯和天花吊灯不谋而合，也增加了空间的宫廷气息。

The basement has a relaxed and casual atmosphere, antique finished cultural stones and logs are used in the billiard room, cellar, wine tasting area, casual dining room and other spaces, which are more accord with the owner's requirement of a relaxed state for leisure life. The facing slab is decorated with cultural stones, presenting an original and vintage flavor. Exposed wood beams on the ceiling show a country style casual feeling, the wood furniture, iron lamp, brown or wood color furniture make the space reveal a unified and harmonious sense.

地下室空间主打轻松休闲氛围，台球室、酒窖、品酒区、休闲餐厅等空间，选用做旧的文化石和原木，更加符合主人休闲娱乐、品位生活所需要的自然轻松状态。文化石装饰的墙面和垭口，原始又复古；外露木梁的天花，充满乡村休闲的逸致；木质家具、铁艺吊灯，褐色或木色家具的选用让地下空间散发统一和谐的休闲设计之感。

Retro Complements with Art
复古与艺术相生

Project Name / 项目名称：Mahwah
Design Company / 设计公司：Diane Durocher Interiors
Designer / 设计师：Diane Durocher
Photographer / 摄影师：Peter Rymwood

In this project, the beige wallpapers connect different rooms, such consistency provides a wide and open visual experience, at the same time, sets an elegant and generous tone. While the embellishment of wine red and deep blue which are rich but not intense makes the space beautiful but quiet. The artistic and historical wood cabinet, romantic patterned carpet and big cultured pictures perfectly achieve the harmony between retro and art. The designer wanted to build such a classical space that is permeated with grace and gorgeousness.

项目以米黄色壁纸串联起不同的居室空间，色彩的连贯性带来持续宽大开阔的视觉感受，同时奠定优雅大方的空间氛围。而点缀在其中浓郁而不热烈的酒红、深蓝等颜色，则烘托出空间华美却又安静的气质。极具艺术感与历史感的实木斗柜、浪漫印花地毯、文化气息浓厚的超大幅挂画，完美实现了复古与艺术的和谐相应。设计师想要营造的，就是这样一个处处洋溢着优雅与华美的古典空间。

The luxurious carpet and pictures show a retro style in the space. A concise and elegant dark blue velvet sofa is selected in order to balance the warm and cold colors. The sofa's intense color reveals a low-key luxury, while its solid structure and soft touch provide a comfortable seating area. The stylish bedroom has a chair in graceful lines, it not only relaxes your body, but also shows your taste and style.

华丽的印花地毯及挂画，客厅空间复古格调可见一斑。而为了平衡客厅的冷暖色调及视觉感，选择设计简洁优雅的深蓝丝绒沙发，浓郁的深蓝流露出低调的华贵，稳固的构造、柔和亲肤的触感，更为此空间增添一份细腻和柔情。别有生活情调的卧室空间，放置的则是线条婉约柔美的躺椅，让你在放松休息的时候也能尽显品位与格调。

The dining room performs colors, movement and layering sense so perfect that it naturally manifests a romantic taste. The water wave valances are connected with the fabrics, creating a sense of rhythm. Wine red seats and shiny gold table not only add radiance into the space, but also increase senses of expansiveness and layering, meanwhile, they bring a strong visual impact, therefore, they are very suitable for a warm but not tacky dining space.

餐厅空间则将色彩、运动感、空间层次感发挥到极致，自然流露出浓郁的浪漫主义情调。水波帘头与帘布的相互结合，富有律动感。酒红色座椅以及金色亮面餐桌为空间增添新鲜色彩，不仅给空间增加了延伸感和层次感，还带有强烈的视觉冲击力，十分适合用于打造热烈而雅致的进餐空间。

Project Name / 项目名称：Longboat Key
Design Company / 设计公司：Todd Richesin Interiors
Designer / 设计师：Todd Richesin
Project Location / 项目地点：Longboat Key, Florida, USA
Area / 面积：578m²
Photographer / 摄影师：Ben Finch

Fragrance in Summer
芬芳夏日长

This warm and remarkable space is filled with a taste of summer, the charming tropical scenery is perfectly highlighted in all details, and different furniture and ornaments are integrated into a vibrant and happy movement, filling every corner. The design team used the colors from the poster collection as inspiration for the design of the condo, so they selected a beautiful wallpaper from Farrow and Ball, their Silvergate Damask in soft pink to cover the walls in most of the home and to provide a rich backdrop for the saturated color palette, creating an unforgettable tropical paradise.

这个热烈而张扬的空间里处处洋溢着夏日的风情，旖旎的热带风光在各个细节中被恰当好处地彰显，各异的家具饰品融合交织为一曲活力而快乐的乐章，充盈在空间的每个角落。设计团队用海报收藏品的颜色作为设计公寓的灵感，选用漂亮的 Farrow & Ball 壁纸，用柔软的粉色 Silvergate Damask 覆盖了房子的大部分墙面，为饱和的调色板提供了丰富的背景，打造了这个让人流连忘返的热带天堂。

If apply bright colors in a large scale, the space would become too dazzling. Contrastive colors can be mixed in order to decorate a place. For example, the rose and light pink cushions and pillows share the same color tone but have various shades, these two fresh colors blend together to form a bright picture. The bright pure cyan velvet sofas are very glittering under the sunlight. All of them present a warm and beautiful summer style.

如果大面积运用明度和亮度都较高的色彩，会使人有眼花缭乱之感。在色彩运用及搭配时可尝试混搭与对比。如玫瑰与浅粉的坐垫与抱枕，即为同一色系的深浅不同搭配，两种高彩度的鲜妍色调融在一起，构成一幅明媚的画面。搭配色泽明亮纯正的青色丝绒沙发，在日光下闪烁出诱人的光泽，营造出热烈又艳丽的夏日风情。

The ceiling color of the dining room corresponds to the background wall, and the carpet color matches very well with the furniture, these two contrasts are striking yet harmonious. The three-dimensional ceiling which has a white undertone is surrounded by golden color to set a gorgeous and noble atmosphere, whilst coordinating with the golden picture frames. The reddish brown dining chairs have classical, soft and exquisite curves that fit in a strong retro style. Reddish brown, tawny and other intense colors with low brightness are collocated with the ornate and bright gold, together they bring an elegant and luxurious ambience.

餐厅的天花色彩与背景墙呼应，地毯色彩与家具相协调，两种对比既有碰撞也显和谐。白色为底金色环绕的立体天花吊顶奠定空间华丽高贵的空间氛围，与金色画框相呼。红褐色餐椅有着古典柔和的玲珑曲线，适合营造浓郁的复古情调。红褐、茶色等低明度浓郁色调搭配着华丽的亮金，缭绕出典雅华丽的氛围。

Project Name / 项目名称：Miami House
Design Company / 设计公司：Frank de Biasi Interiors, LLC
Designer / 设计师：Frank de Biasi
Project Location / 项目地点：Miami, USA
Photographer / 摄影师：Mark Roskams

A Beautiful Mix of Colors
斑斓色彩汇

The owner wanted to have a relaxed and unconventional living environment, so the designer created a house according to this request. The selection of different colors aims to not only create a certain style, but also make a more pleasing impression, meanwhile, special ornaments are used to coordinate the whole space, so that a mix of different items produces a wonderful sense of harmony. A desired effect like this requires bold attempts and repeated experiments.

屋主追求轻松且不落窠臼的居室环境，设计师充分考虑到屋主的诉求，把这份舒适生活的诉求付诸实践，便有了空间的设计展现。设计师选择好几种不同的居室色彩不仅仅是为了某种风格服务，更是为了呈现出一个更加赏心悦目的空间印象，再用特色的装饰品来协调整体，使各异的物品摆放在一起产生奇妙的和谐感。这需要大胆的尝试和反复试验才能达到理想中的效果。

177

Colors of the dining room are so bold that they form a dramatic contrast. The wall with a window is painted with warm red, while the window frame is specially decorated with yellow, it is this strong contrast that reveals an inexplicable charm. The cyan round table surprisingly balances the color tension and stabilizes the atmosphere of the entire space.

餐厅在色彩的运用上极为大胆，充满戏剧性的对比。开窗一面的墙壁由热烈的红色染就，而在窗框部分则别出心裁以黄色装饰，这种强烈的对比却有着一种道不出的魔力。而青色圆桌则奇异地平衡整个空间喷薄欲出的色彩张力，稳定了空间内的气氛。

Most of the colors in the bedroom share a same tone. Starting from the corridor, deep or shallow cyan covers a large area. Fresh and elegant blue white bedding looks clean and bright, such pure and clean color makes the soul feel quiet, just like the feeling given by the vast sky, and it's suitable for creating a tranquil and comfortable living space. The appropriate mix of lavender headboard cleverly enlivens the space.

卧室空间多同色系的铺垫,由廊道开始,或深或浅的青色铺设了大半空间。清爽雅致的蓝白床品,显得纯净明媚,仿佛高远的天空般,纯粹干净的色调让人的心灵也宁静起来,适合打造宁静舒适的居室空间。适当加入淡紫色床头架,巧妙活泼空间氛围。

When the furniture shape and lines tend to be simple and practical, the colors can be more flexible. Pink walls set a warm and fresh tone for the entire space, and the mild yellow sofas achieve a comfortable and bright atmosphere. The sofas and pillows are decorated with patterns of large blooming flowers, making the space naturally bright and extraordinarily beautiful.

当家具的造型与线条趋于简洁实用，空间的色彩就有了更多发挥的空间。粉色的墙面为整个空间铺垫一种温暖清新的气息，而沙发换上暖暖的浅黄色调，成就舒心明亮的空间氛围。以大朵盛开的妍丽花卉图案装饰沙发及抱枕，空间自然明媚起来，透出不同凡响的美丽。

Inspiration of a Beautiful Home
家的灵感与美丽

Project Name / 项目名称：St. Regis Bal Halbour
Design Company / 设计公司：Geoffrey Bradfield, Inc.
Designer / 设计师：Geoffrey Bradfield
Project Location / 项目地点：Paris, France

For habitants, "home" is not a concept, but an experience, a "true" dialogue with themselves and with the world. Modern people no longer require swanky, tedious or showy symbols to highlight their status, they are confident about their own aesthetic sense, they don't blindly follow the trend, but care about the quality of life so much that they pay more attention to the inner spiritual world and needs. In this project, the designer presented us a lively, elegant and stylish home from inside.

对居住的人而言，"家"不是一个概念，而是一种体验，一种"不忘初心"的与自己、与世界对话。现代人对于家的追求，不再需要那些张扬的、繁琐的、炫耀式的符号来凸显自己的地位，他们对自己的审美充满自信，不盲目追随潮流，对生活品质极致考究，从而更加关注内在的精神世界及需求。本案设计师从内在出发，给我们诠释一个活灵活现的极尽优雅格调的家。

Facing the sea, you can be a painter and play with various colors. The designer brought natural sea blue into the space, and used it as the main tone. Collections with certain themes can increase the quality of the design and make yourself feel delighted. The simple colors are well matched with blue, creating a visual balance.

面朝大海，做个画家，玩转色彩。将天然的大海蓝，注入整个空间，成为色彩的主旋律。一定的主题元素的收藏品，会把家的设计带进专业级别，令自己愉悦。同时色彩简约，与蓝色相衬，可以创造一种视觉的平衡感。

A beautiful home is created not just for an eye-catching effect, what matters more is its comfortable feeling. You can feel the high texture through the touch of natural materials, such as silks, satins and cotton linens. Because of more expectations in life, home is no only just a house, but a stage for everyone who loves life.

家的美不仅为了养眼，身体的舒适才是至关重要。用双手去触碰那些绸缎、棉麻等自然材质，让家的质感更显高级。因为对生活有了更多的憧憬，家，已不仅仅是一座房子，它甚至可以被搬上舞台，带给每一个热爱生活的人们。

Design Companies / 设计公司：Shearly Investments, Ltd. & Patricia Reid Baquero Diseño de Interiores
Architects / 建筑师：Juan Mubarak, Alejandro Marranzini, Cristina Pérez
Preservation Architects / 保存建筑师：Jose Batlle, Danae
Interior Designer / 室内设计师：Patricia Reid Peguero
Project Location / 项目地点：Santo Domingo, Dominican Republic
Area / 面积：508m^2
Photographer / 摄影师：Tiziano de Stefano

An Elegant World in Colonial Style
殖民风格中的文雅世界

Casa del Árbol is located close to Parque Duarte, Convento de los Dominicos. It is a typical XVI Century colonial corner house where the rooms in the perimeter are aligned with the street, forming corridors that surround a charming interior courtyard. Roman roofs, wells, brick arcades, and masonry walls have been conserved, becoming the basis for the design of eclectic spaces that combine the antique with the contemporary. The interior design showcases Dominican culture utilizing local art pieces, carved modern furniture and antique tropical pieces. Each of the houses is designed with a perfect balance between the historic and the contemporary, and decorated with different themes referencing their history, generating diverse environments within the Casas del XVI collection.

Casa del Árbol位于多米尼加修道院杜阿尔特公园附近，是一个典型的16世纪殖民风格的房子。房间围绕林荫小道排列，形成一个环境迷人的内部庭院走廊。罗马屋顶、水井、砖石拱廊和砖墙都被完整保存了下来，整个住宅是古典与现代风格相结合的折衷设计。室内设计则通过当地艺术作品、雕刻的现代家具和古老的热带艺术品等展示了多米尼加文化。每一间房屋的设计都寻找历史和当代之间完美的平衡，并以不同的主题来装饰定位它们的历史，这些在16世纪住宅系列中产生不同环境的美感。

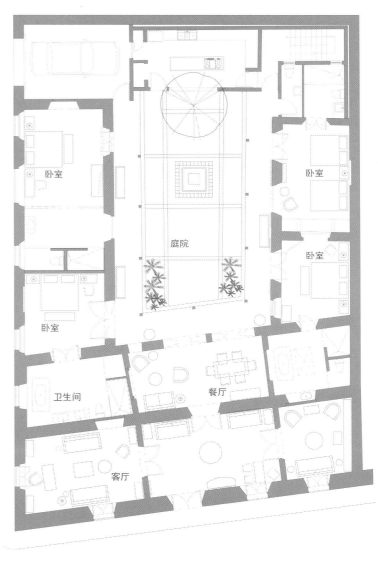

Decorated around the theme of the pineapple, a symbol of hospitality, interior designer Patricia Read achieved a delicate décor, in a lovely ambiance. The sitting room is a cozy little room. Architect William Reid designed the large iron gratings gracing the walls in the 1970's and they hang next to a long African ceremonial mask. Other items in the room are works by the artist Henriquez, a current resident of the Colonial City.

In the Library, comfortable navy blue sofas, antique armchairs and a mix of cultures and art welcome the guests and provide them an artistic and cozy space. On the walls, antique architecture prints are juxtaposed with paintings on palm tree slabs made by local artisans. A large handmade tapestry adorns one of the walls, contrasting the brick-colored walls. In the main niche chubby cherubim sculptures fly, as if echoing the ornate altars of the many colonial churches nearby, creating an interesting space that is full of solemn and sacred senses.

围绕着象征好客的菠萝主题装饰，室内设计师 Patricia Read 在令人愉快的氛围中创造了一种精致的装饰之美。虽然客厅是一个舒适的小房间，建筑师 William Reid 在 20 世纪 70 年代围绕它设计了一个大铁栅栏来优化墙面，而他们把它挂在一个长长的非洲面具旁边。房间里的其他摆件都是艺术家 Henriquez 的作品，他是这个殖民地城市的居民。

藏书室里，舒适的海蓝色沙发、古董扶手椅以及各种文物和艺术的交织欢迎着客人的到来，为客人们展示一个充满文艺气息又不失惬意的空间。墙壁上古式的建筑版画与当地工匠制作的棕榈树壁画并列摆放，巨大的手工挂毯与砖墙形成鲜明的对比。主壁龛上胖乎乎的小天使雕塑，仿佛在呼应附近许多殖民地教堂的华丽祭坛，饶有趣味又充满了庄严的神圣感。

Exotic Influence
异域来风

Project Name / 项目名称：Italian Villa
Architectural Company / 建筑公司：Landry Design Group
Interior Design Company / 室内设计公司：Joan Behnke & Associates
Architect / 建筑师：Richard Landry
Interior Designer / 室内设计师：Joan Behnke
Project Location / 项目地点：Los Angeles, USA
Area / 面积：1970m^2
Photographer / 摄影师：Erhard Pfeiffer
Main Materials / 主要材料：Marbles, woodworks, chandeliers, decorative pictures, etc.

The design team spent three years building this dream home for the owners. They traveled extensively and are very fond of different design styles in Europe, therefore, not only the European architectural elements were used in the exteriors, but the Art Deco decorative arts became part of the fabric of their home. From cabinets created from exotic woods like rosewood, makassar ebony, to a Ruhlmann inspired fireplace mantle that had graced a Parisian apartment, to striking light fixtures made by the historic Lalique company, such vibrant colors and rich textures are very impressive. The team developed them to a personalized design vocabulary and created a unique living space. Being here, you will find it's wonderfully unusual.

设计团队花费了三年时间为屋主量身定制了他们的梦之家。屋主喜欢旅行，非常中意欧洲各国设计风格，所以不仅在外观的建造上借鉴了欧洲的建筑元素，室内装饰上也处处可见欧洲各国艺术饰品的痕迹。蔷薇木、檀木打造的古木时代的橱柜，鲁尔曼风格的壁炉，在历史悠久的莱丽卡公司生产的灯具，鲜明的色彩及非凡的质感让人深深着迷。设计团队将他们开发成一个个性化的设计语汇表，打造出这个独一无二的居家空间。遨游其中，你会发现它与众不同的魅力。

At the beginning of the project, designers wanted to present a "design experience" for the owner by taking them on several exploratory shopping trips to Italy, France, and New York in anticipation of finding one-of-a-kind pieces that related to the Art Deco period. Stone was resourced from Italian quarries, unique Deco inspired millwork for powder rooms was built in France, lighting was custom fabricated in the suburbs of Paris, and art was collected internationally.

设计之初，设计团队希望为屋主呈现一种"设计体验"。从他们在意大利、法国和纽约等探索性购物中，发现一些装饰艺术时期的特色作品，并将他们应用在设计中：石材源自意大利的采石场，化妆间独特的装饰艺术风木质工艺制作于法国，照明设备是在巴黎郊区定制的，而艺术藏品则来自于世界各地。

To accommodate the owners' frequent entertaining and hosting of philanthropic events throughout the year, the functionality of the site was maximized by designing a full basement with a parking garage, indoor swimming pool, movie theater and game room, as well as a first floor with distinctive spaces like dining room, guest room and living room. Moreover, the unique geographical advantage on the outside contributes to a pool and large garden space. Every visitor would feel at home when chatting on the corridor or dancing on the lawn.

为了满足屋主们全年的娱乐活动及举办慈善活动的需求，在地下室配有停车库、室内游泳池、影院和游戏室，地上一层则设计了各具特点的餐厅、会客室、客厅等空间，为待客及举办宴会提供了充裕的场地。室外更是借助得天独厚的地理优势，配备有露天游泳池及超大面积的花园，无论是在廊上闲坐聊天，还是在草坪上翩翩起舞，都会让每个来访者有宾至如归的感觉。

A Luxury Home Filled with Happiness
幸福洋溢的奢华之家

Project Name / 项目名称：Old Greenwich Residence
Design Company / 设计公司：Marks & Frantz Interior Design
Designers / 设计师：Lydia Marks, Lisa Frantz
Project Location / 项目地点：CT, USA

This is the second house that Marks & Frantz have renovated and designed for this young couple. Shortly after the birth of their first child in 2010, the couple moved from their Flatiron district loft in New York City to raise their child in Greenwich. The house they purchased was a 1920s Tudor with great bones, but it felt dark and claustrophobic. The clients searched for the designers responsible for the interior decoration of the Sex and the City movies and found Lydia Marks & Lisa Frantz, who opened up the rooms, added larger windows and modernized that house to suit the young couple. They also added a pool and beautifully landscaped terraces and opened up the back of the house to realize the home's potential for indoor/outdoor living.

这是设计师为这对年轻夫妇业主设计的第二套房子。2010年,他们的第一个孩子出生后不久,这对夫妇从位于纽约市熨斗区的阁楼搬到了格林威治抚养他们的孩子。他们买下了一个20世纪20年代铎式风格的大房子,但它给人一种黑暗和幽闭的恐怖感。客户想找负责《欲望都市》电影室内装饰的设计师帮他们设计房子,最终找到了Lydia Marks和Lisa Frantz。设计师帮他们扩展了空间,增加了更大的窗户,并让房子看起来更具现代化以匹配这对年轻夫妇。同时,还增加了一个水池和漂亮的景观露台,开辟利用屋后的空间,以实现室内与室外生活的更多可能性。

When the couple became pregnant with their second child, they decided that they needed more room and found a house in Old Greenwich on a sleepy street with a small-town feel. They once again called on Marks & Frantz to give the home a special feeling. Although the house, built in 2000, had some beautiful architecture and nicely proportioned rooms, it also had a mix of styles that were not appropriate to the house, including a sunken living room.

当这对夫妇怀上第二个孩子的时候，他们觉得需要更多的房间，就在老格林威治的一个小镇的安静街道上找到了一间房子。他们再次找到设计师 Marks 和 Frantz，希望给家里营造一种特殊的感觉。尽管这栋房子建于 2000 年，房间却有着漂亮的建筑结构和恰当的比例。只是有一些设计不适合房子的风格，如一间下沉式的客厅。

The designers' goal was to pay homage to the traditional style of this house while making it feel young and fresh. The duo's knack for redefined traditional décor is a hallmark of their practice, so they set about modernizing the kitchens and bathrooms, raising the sunken living room and enhancing the architectural details to tell a more complete story. They filled the house with extraordinary details, fixtures and finishes, and many pieces of furniture they custom-designed for the space. This couple love to entertain both informally and formally, and Marks & Frantz were able to bridge the gap with the design so that the rooms look elegant yet still feel inviting and warm.

设计师的目标是要向传统风格致敬，同时让客户感到年轻与清新。他们的设计以重新定义传统装饰风格的手法为特点，因此他们改善了厨房和卫生间，元素更现代化，升高客厅，并强化建筑细节以讲述一个更完整的故事。另外，在房子里布置了一些特别的细节、固定装置和装饰，为空间定制了许多的家具。为了满足这对夫妇对非正式和正式娱乐的喜爱，Marks 和 Frantz 弥合了设计之间的差距，使房间看起来优雅、充满魅力且温馨。

223

Extreme Design Interprets a Remarkable Temperament
极致设计 演绎非凡气质

Design Company / 设计公司：SC Editions
Designer / 设计师：Stephanie Coutas
Project Location / 项目地点：Paris, France
Area / 面积：365m²
Photographer / 摄影师：Francis Amiand

Throughout the project, Stephanie's "savoir faire" comes to show with plenty of luxurious details, while maintaining a settled and contemporary feeling. The renovation has taken 18 months of hard work, with the aim to highlight her impressive and extensive art collection that she has accumulated through the years. The agency's main target is to develop unique and exclusive projects that exceed the expectations of her international clientele. The result is a warm and cozy yet spectacular flat.

在整个项目中，设计师 Stephanie 的"才能"凸显在大量奢华的细节上，同时保持了沉稳的现代感。这次翻修花了 18 个月的时间，目的是为了突出她多年来积累的令人印象深刻的大量艺术收藏。设计公司的主要目标是设计一个超出其国际客户预期的独特项目，他们最终打造了一个温暖舒适而又壮观的住宅。

Starting from the entrance in white opal with brass inserts to the motif of the ceiling going through the reception hall, the combination of classic moulding styles, revisited through a modern pattern gives a new sense to the place. The main idea was to create a fluid, spacious hallway that merged with the reception, creating the sense of an art gallery.

从有黄铜嵌入天花板图案的白色蛋白石入口到接待厅，与经典的造型风格结合，经过现代形式重新规划，赋予了这个空间新的意义。设计师的主要想法是创造一个流畅的、宽敞的走廊，与接待厅融合在一起，营造出一种艺术画廊的感觉。

As always, the design of the house brings feelings of comfort and peace to the soul. All the bedrooms are ensuites and have all the luxuries of 5 star palaces such as mini bars, high quality beds and linen, Mr and Mrs separate bathrooms and dressing areas. Materials and finishes are classic and elegant in tones of black and white with bronze and brass finishes, creating a fine, luxurious and modern home.

房子的设计是希望给灵魂带来舒适与安宁。所有的卧室都是套房，拥有 5 星级宫殿的所有奢侈品，如迷你酒吧、高质量的床和亚麻布、独立的男女浴室及化妆间等。材料及涂饰也都是经典而优雅的黑色、白色的青铜或黄铜饰面，打造成一座精致、奢华、摩登的极致家。

A Second Friendly Nature
第二个友好的自然

Design Company / 设计公司：Marcel Wanders Studio
Designer / 设计师：Marcel Wanders
Project Location / 项目地点：Amsterdam, Netherlands
Area / 面积：240m²

As a product and (latterly) hotel designer, Marcel Wanders needs no introduction, but the domestic interior is a new departure for him. The apartment samples the rare and exotic – in this case, the currently endangered crafts of Dutch plaster molders and parquet layers and Indonesian wood carvers, alongside specially selected and custom-made items of furniture. The visual richness is controlled by a certain sense of proportion and an orderly orchestration of the limited space: barely an inch is wasted, or left undecorated.

Marcel Wanders 是一个闻名遐迩的产品设计师和酒店设计师，但家装设计对他来说是一个新的起点。本案是一个稀有样品住宅，独具异国情调。如目前濒临消失的荷兰石膏模型、石膏板层和印度尼西亚木雕，以及特别挑选和定制的家具。视觉的丰富性是由一定的比例和有限空间的有序排列所控制：每一寸空间都被利用和装饰。

The garden and the house had to live together in a beautiful way… The designer wanted to blur the boundary between inside and outside, and between architecture and nature. So the garden has a marble floor, which takes the interior into the exterior, and it has concrete topiary, and even the real plants – and they're only the vines covering the walls – are dotted with porcelain flowers that are also a form of lighting. We are artificial beings, that's our nature. Cows eat grass, birds fly, and humans make things. We call that artificial, but it comes so naturally to us. A few years ago, Andrea Branzi said that I interpret technology as 'a second friendly nature'. I think that about sums it up.

花园和房子一起以美妙的方式构成了生活。设计师想要模糊内部和外部之间以及建筑和自然之间的界线。因此花园内有一层大理石地面，将室内装饰延伸至外部，还有混泥土树木造型和真正的植物——覆盖在墙上的藤蔓，它们点缀着起照明作用的陶瓷花朵。我们是人为的生物，那是我们的本质。牛吃草，鸟会飞，人类创造东西，我们称这是人为的，但这对我们来说是自然而然的事实。几年前，Andrea Branzi 将科技诠释为"第二个友好的自然"，我觉得这概况了一切。

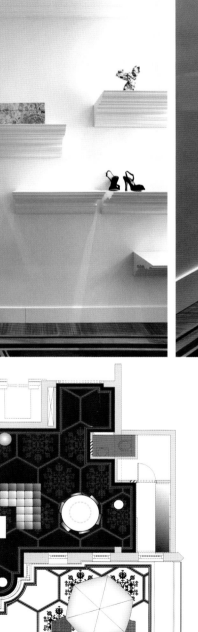
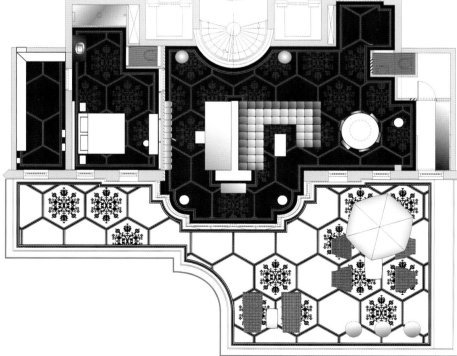

We hid most of the functional stuff – the cables, the radiators, the sockets. So that allows the architecture to speak. And we gave the space a certain monumentality, by enhancing the semi-circular wall in the main space (it contains the staircase). We made the curved wall bulge to give it an iconic presence. And then we added three concave columns that echo that shape. We also got rid of most of the doors, because doors would stop you understanding the space. So the kitchen is behind an etched steel wall, not a door.

Although the design is very detailed, it's restful in other ways. For example, the repetition of the hexagonal grid for both floor and ceiling, and the molding running around the top and bottom of the walls. The client gave us carte blanche. Actually, when she saw the finished result, her response was, 'I would never have done any of this myself. But I love it!' Of course we'd started off by talking about her preferences and we took direction from these in a general way. She likes delicate things, richness, elaboration, lots of fantasy. But on the other hand she lives in a casual way. And finally, we gained her praise.

我们的设计中隐藏了大部分的功能物品——电缆、散热器、插座等，这样就更好地突出建筑结构。通过强化主空间的半圆形墙（包含楼梯），给了空间一个特定的纪念性。同时把弧形的墙凸出来展示其标志性的形象，再增加三个凹进的圆柱以呼应这个形状。另外设计时我们移除了大部分的门，因为门会阻止你理解空间，所以厨房的前面设计了蚀刻的钢制墙面，而不是门。

虽然设计非常精细，但是仍给人轻松的感觉。例如，六边形网格重复出现在地板和天花板上以及在墙壁顶部和底部周围的模塑上。设计方面，客户让我们全权负责。当她看到最终的结果时，她说："我自己从来没有做过这样的事情，但是我爱它！"整个项目我们从谈论她的喜好开始，再从整体的角度出发，以她喜欢的精致、丰富、用心、爱幻想为设计原点，扩散到她喜欢的随意的生活方式，最终获得客户的赞誉。

Design Company / 设计公司：Sara Story Design
Designer / 设计师：Sara Story
Project Location / 项目地点：Bin Tong Park, Singapore
Area / 面积：1208m²
Photographer / 摄影师：Masano Kawana

Customize the Home with Art
定制家的嫁衣

The design objective in creating this residence was inspired, in part by its geographical locale, combining the client's extensive Asian art collection with vintage and contemporary design creating a truly unique and elegant space.

The open, double-height living room maximizes space and natural light with large windows and streamlined architecture inspired by the Raffles Hotel in Singapore. The space combines pops of blue as seen in the custom 'Snorkel Blue' rug, antique chairs and combined with neutrals shades of cream and gray. The dining room is dramatic and sophisticated inspired by The Peacock Room by architect Thomas Jeckyll as seen in the custom millwork displaying the client's extensive porcelain collection. The cabinetry beneath is outfitted in Meriguet embossed goatskin panels in a traditional Asian fan motif. The space combines an antique 1950's Italian Glass Cabinet, Murano Glass Wall Sconces with a custom walnut and brass dining table. The room combines antique furniture and lighting – like the Swedish brass floor lamp (c. 1960) with Serge Mouille Applique Sconces, Art Deco Side Tables, four-panel antique Asian Screen and large porcelain urns. The tall and elegant space is wrapped with art.

这个住宅的设计部分是受其"地理区域"的启发，将客户收藏的大量亚洲艺术藏品与当代设计相结合，创造出一个真正独特而优雅的空间。

开放式的双层高客厅将空间最大化，引入自然光线的大窗户与流线型的建筑风格的灵感来源于新加坡的莱佛士酒店。空间融合了定制的深海蓝色地毯，古色古香的椅子，以及中性的奶油色和灰色的沙发。同时，定制的打磨工艺可以看出餐厅的设计很精细且引人注目，设计灵感来自建筑师Thomas Jeckyll设计的孔雀之屋，室内装饰充分展示了客户的瓷器收藏。橱柜上镶嵌着带传统亚洲图案的凹凸山羊皮面板。这个空间将20世纪50年代的意大利玻璃橱柜、慕拉诺玻璃墙和定制的胡桃木及黄铜餐桌结合在一起。此外，房间还摆放着古董家具，包括瑞典的黄铜落地灯、Serge Mouille嵌花壁灯、装饰艺术风格的边桌、四格的古式亚洲屏风和大陶瓷瓮等。将如此高挑、素雅的空间，裹上艺术的嫁衣。

The master suite combines a custom wood and leather king size bed with Parchment Bedside Tables. The seating area combines a pair of custom lounge chairs and a 1940's Breche Marble Coffee Table by Jean-Philippe Demeyer, which are very convenient for communication and daily use. The client's dressing room is encased in floor-to-ceiling antique glass panels resembling a jewelry box. The 1940's black lacquer antique vanity and chair complement the space perfectly.

主卧套间里有定制的木质大号床和羊皮纸床头柜。在座位区有一对定制的躺椅和一张1940年代由Jean-Philippe Demeyer 设计的 Breche 大理石咖啡桌，为主人之间的畅聊与生活提供极大的便利。客户的更衣室被围合在地板和天花板之间的老式玻璃面板中，像一个珠宝盒，而20世纪40年代的黑漆古风梳妆台和椅子则让空间显得更加贴心完美。

A guest bedroom is located on the first floor and also serves as a library. The room is wrapped in a hand-painted aquatic design wallpaper with Koi fish and lily pads. There is an antique bed from Thailand and a seating area with antique side chairs and Gio Ponti table. Hanging above the bed is an Arne Jacobsen glass globe by Louis Poulsen. There are three kid's rooms on the second floor – two girls and one boy. The rooms incorporate each child's favorite color – and each is reflective of their personalities.

位于一楼的客房，也可用作图书馆。这个房间墙壁上铺陈着手绘的锦鲤鱼和睡莲的水彩画墙纸，以及从泰国买的古董床等。另外特设了一个座位区，座位区里面摆放着古雅的边椅和 Gio Ponti 桌子。床上方悬挂着 Louis Poulsen 公司的 Arne Jacobsen 玻璃球。沿阶而上，二楼有三个儿童房，其中两个女孩房和一个男孩房。这些房间都融合了每个孩子最喜欢的色彩，反映的都是他们自己的个性。

Project Name / 项目名称：A Bexley Estate
Design Company / 设计公司：Geoffrey Bradfield, Inc.
Designer / 设计师：Geoffrey Bradfield
Project Locatoin / 项目地点：Bexley, London, UK

Fresh and Delightful Life

清新愉悦的惬意生活

Style of this estate is not classical or modern, but something between the two. Here you can see the green lawn and leisure area where the owners can enjoy a beautiful life every day. The villa retains some classical and traditional elements, such as Roman columns, marble fireplace, cultural bricks, arch porch, detailed sculptures and others, together they build an image of time and space. Meanwhile, it has a free and bold modern color scheme which impresses people in each space. In terms of furniture, all are well crafted with clear lines, while some of the irony ornaments bring us to different ages.

这是一座室内设计介于古典和现代的庄园别墅，庄园中有绿意盎然的草坪和惬意的休闲区，美好的体验带给生活每天享受的喜悦。古典在于它保留了传统元素，罗马柱、大理石壁炉、文化砖、拱形门廊、细节的雕刻等，并以此塑造了空间沉淀岁月的形象。同时它的现代在于色彩搭配自由、大胆，每个空间中总能找到让人眼前一亮的配色方案。家具的选择也是线条清晰又讲究工艺，一些金属铁艺摆件也传达出时代的气息。

In the square dining room, the round glass table is particularly bright and clean, and well matched with the transparent glass tableware. The gray walls and gray purple dining chairs complement the space, in which the pure colors are harmonious and unified. However, the abstract green oil painting on the wall and the red cabinet below it form the most striking combination, which makes a colorful, lively and vivid space.

方形餐厅中选择圆形玻璃餐桌尤为明净，与透明的玻璃餐具也很匹配。而灰色的墙面和灰色偏紫的餐椅是对透明餐桌的补充，纯正的色彩和谐而又统一。但最巧妙的是墙面挂着一幅绿色抽象油画和底下红色的矮柜，给空间一道绚丽色彩，显得活泼生动。

Compared to other video rooms, this room also chooses dark color as the background to create a good visual contrast effect. But what makes it different is that its design refers to the arrangement of a sitting room. There are warm fireplace, randomly placed sofas and table, as well as eye-catching pictures, at the same time, the mirrors not only visually enlarge the space, but also provide a more open, comfortable and enjoyable viewing experience.

与其他影音室设计相比,这间影音室同样选择深色为室内背景色,为观影创造一个良好的视觉对比效果。但它的不同之处在于以客厅摆设形式来参考设计这间影音室,温暖的壁炉,自由的沙发桌几的摆放,醒目的装饰挂画,同时用镜面装饰来扩大视觉空间,给人以更开阔和舒适的空间来享受观影。

In the bright and spacious master bedroom, the bedding and curtain are made of light color fabrics, so that an integrated atmosphere is created yet each space is clearly divided. The light fabrics bring a comfortable and relaxed feeling, and are well suited in a bedroom which is open and close to the green mountain. The carpet with a light blue undertone is like the blue sky after rain, revealing an easy, delightful and cozy ambience, which is just what the room needs. Several doors and windows lead to the balcony, which enables the habitants to enjoy the natural scenery closely, breath the fresh air, as well as relax their minds and bodies.

在明净宽敞的主卧房里，床品和窗帘选择同样的浅色布艺，让空间一体化但又各自保持距离，浅色布艺带来的舒适和清爽也更加适合这间开阔并且临近无限绿意山景的卧室。以浅蓝色为底的地毯如雨过天晴后的蓝天，散发出令人轻松的喜悦，这恰好是卧室所需要的惬意氛围。多扇门窗通向阳台，近距离接触室外自然风景，呼吸清新空气，可以扫去身心疲惫的倦意。

A Young House that Releases Personality
释放个性的青春家

Design Company / 设计公司：Gabriele Pizzale Design Inc.
Designers / 设计师：Gabriele Pizzale, Melissa Gorton, Zaynab Shirazie
Project Location / 项目地点：Mississauga, Ontario, Canada
Area / 面积：743m²
Photographer / 摄影师：Mike Chajecki

This home was custom built for a young family that loves colour! The clients were very adventurous and gave designers free reign for creativity, they wanted a home that they could entertain in.

The most important mandate was to create a wow factor! They wanted colour and unique pieces that were functional as well as beautiful. The large home needed pieces that were the correct scale, and special enough not be to over taken by the vastness of the rooms. Many custom pieces were utilized to make sure this home was unique and one of a kind. From the moment you enter the home you are enveloped with luxury and vibrant hues. The millwork on the walls lends an air of sophistication, and the unique console with the shell inlay is stunning beside the custom built chairs. The use of contrasting textures like the Tibetan fur on the stool under the sleek inlay walnut console, is particularly striking in the family room. Bold use of orange and purple on the upholstery and drapery adds a vibrance to this space.

这个家是为一个喜欢色彩的年轻家庭定制的。他们都很有冒险精神，给了设计师自由创造的机会，希望设计师打造一个可以缤纷娱乐的家。

屋主给设计师最重要的任务是创造一个令人惊叹的家，希望色彩独特，装饰品要美观又实用，同时大房子需要尺寸合适，此外，需用个性的装饰品来匹配房间的宽阔。因此设计师通过独家定制，定制了许多饰品以确保这个家是独一无二的。从你进入门厅的那一刻起，你就会被充满奢华与活力的色调所包围。墙壁上的打磨工艺营造精致的氛围，而独特的控制台上镶嵌着贝壳，搭配定制的椅子，令人眼花缭乱。位于光滑的胡桃木桌下面的凳子，使用不同质地的纹理，如藏毛，在客厅里显得特别醒目。而色彩选择上大胆地使用橙色和紫色，给空间增添了活力与热情。

The dark plum walls and extensive dark stained oak table really speak to this the elegance of the dining room. The dining chairs have sexy curves, that its colorful backs play off of the white color, orderly arranged and exquisite.

深色的玫红色墙壁与宽大的深色橡木桌，增添了餐厅的优雅。摆放着的印花线条餐椅，性感的曲线在椅背上延伸，与白色的内里形成良好的互补，一字排开，十分别致。

The basement is all about entertaining, with a lounge like sitting area, and the swivel purple leather chairs. The bar area has wood slats wall treatment and mosaic tile. The brown leather chairs in the theatre room fully recline for an ultimate movie watching experience. The second floor is much more subdued in colour, these are the private spaces for the clients. The master bedroom is made up of soft hues, with a soft blue ceiling and white bedding, great for relaxing.

地下室全是娱乐性的空间,休息室有旋转的紫色皮椅。酒吧区有木板墙和马赛克瓷砖。影音室里的棕色皮椅可以倾斜调整到最佳观影状态。二楼的颜色则柔和许多,且注重屋主的私密性。主卧室由浅色调组成,有柔和的蓝色天花板、米色的墙壁、白色的床品等,适合放松与休憩。

Slow Life in a Rural Home
田园之家的慢生活

Design Company / 设计公司：VSP Interiors
Designer / 设计师：Henriette von Stockhausen
Project Location / 项目地点：England
Photographer / 摄影师：Alexander James

This house is filled with a touch of rustic style, and its interior decoration combines modern European style with rural taste, in this way, it meets the requirements of modern living and allows people to enjoy a slow life. Inside the house, lots of natural and pastoral wood materials are applied, such as the long table, drawer chest, sofa chairs, bed and exposed ceiling, all present a natural and plain temperament. If rural style is the best choice for slow life, then a modern tone gives the house a new appearance. Ornaments finished by modern craft and colors contribute to an artistic space.

这间住宅充满了淡淡的乡村风格的休闲气息，室内设计是现代欧式与乡村风格的结合，打造了一个既能过慢生活又符合现代居住要求的住宅。室内多选用靠近自然、具有田园风情的原木材料，如长条的桌子、斗柜、沙发椅、床和外露天花等，都是用原木打造的家具，散发出原木的自然纯朴。如果说田园风格是回归慢生活的最好选择，那现代风格的影子则赋予这座住宅新的观感，现代工艺的摆件和现代方案的配色则碰撞出艺术的交织。

The bedroom uses white as the undertone, while the floor has a mild wood color. The white bedding cleverly responds to the walls, creating a unified sense. The striking part is the ceiling with wooden frame, although the sunken design reduces the height of the space, but it echoes with the floor and forms a rural style layout that is casual and fresh, so that the owner can feel relaxed and comfortable from inside to outside.

卧室以白色的墙面为底色，木色的地面温润，白色床品和墙面巧妙呼应，显示出空间的统一性。而最值得品味的则是木架设计的天花，虽然下沉的天花设计降低了空间的高度，但却与地面相呼应，也散发出田园风格的休闲和清爽的格局，给在此休息的主人从内到外的轻松与惬意。

Although the kid's room is not large, but it's interesting and naive. On the wooden bed, there are cotton and linen lattice products which correspond to the curtain, besides, the stuffed toys indicate the boy's innocent and lively characteristics. The little boy is curious about the natural world, so we can see a painting of insects on the wall, and the white frame on the ceiling looks like a home for crocodile doll, revealing his loving heart and goodwill.

儿童房虽然空间不大，却充满了童趣和童真。木质床以棉麻的格子床品相搭配，与窗帘相呼应，还放有小主人喜欢的毛绒玩具，透露出男孩的天真活泼。小男孩对大自然世界充满了好奇，墙面上挂着昆虫图画，天花板上白色的隔层架像是鳄鱼玩偶的家，也显示出小主人的爱心和善意。

Design Company / 设计公司：S.E. Design Services
Designer / 设计师：Sandra Espinet
Project Location / 项目地点：San Jose del Cabo, Mexico
Area / 面积：1115m²
Photographer / 摄影师：Hector Velasco Fazio

A Resort with Spectacular Sea View
辽阔海景 度假胜地

This project was a labor of love from my client who studies architecture and is a huge lover of style and design. We worked closely together to create this unique one of a kind home and make it as stylish and fun as possible. Each space was studied carefully to ensure its location and design suited his needs. The main concept for the home was usability, style and a casual elegance that reflected his way of living and entertaining.

这个项目的屋主热衷于研究建筑学，是风格和设计的狂热爱好者。我们紧密合作，创造了一个独一无二的家庭，让它尽可能的时髦和有趣。每个空间都经过了仔细研究以确保其位置和设计符合屋主的需要。一个住宅首先要考虑的是可用性，设计风格和随意的优雅，都反映了屋主的生活方式和娱乐方式。

The home tried to be natural and assimilate as much as possible into the natural desert mountain setting and even the landscaping was considered and we used local native plants and cactus so that they blended in. Despite the perfection and the amazing interiors, this home is quiet and not ostentatious at all. It has a quiet casual elegance about it and the fun spirit of the homeowner shines in every room.

在自然的沙漠和山地环境中，屋子给人自然清新的感觉，室内装饰与四周背景相互融合，我们选用当地的植物和仙人掌与自然景观相呼应。尽管有完美和令人惊叹的内部装饰，这个家却安静而不浮夸。它有一种安静而随意的优雅，屋主的趣味精神洋溢在每个房间中。

The first portion of design for this project was work closely with the builder and select all interior finishes and assist with the design of key carpentry areas such as the kitchen and bathrooms. The kitchen ceiling has a traditional Mexican brick dome and is contrasted with transitional carpentry and custom hand made tiles, creating a harmonious cooking space. The second phase was furniture design, layouts and the search for unique custom items we would use to make this home special. Particular attention was made to lighting and we selected huge and very dramatic lights that filled the space as well as created moods at night. Muted hand blown glass and iron were used in our lighting selections.

项目的第一部分设计我们与建造者紧密合作，一起选择所有室内饰品，并协助设计主要的木工区域，如厨房和卫生间。厨房天花板是传统的墨西哥砖屋顶，与过渡期的木工和定制的手工瓷砖形成对比，以此构成和谐的味蕾空间。第二阶段是家具设计、布局和寻找能使这个屋子更特别地独特定制物品。我们特别注重照明设计，选择了巨大的非常引人注目的吊灯，主打无色的玻璃和铁艺，灯光不仅照亮了空间，也营造了夜间的浪漫氛围。

The home sits on the top of a desert mountain and overlooks the Sea of Cortez from above. To blend in with that we selected natural burnt tones of sunset colors to use throughout. We used masculine, large pieces of furniture and unique ceramic art and antique pieces of wood as art, so as to decorate daily life. Open concept floor plan for the great room was implemented and total privacy for bedrooms. In order to achieve the client's expectation, the project took almost 2 years to complete.

这个家坐落在一座荒山上，俯瞰着科特兹海。为与此相融合，我们室内色彩选择了自然的日落颜色为主。同时，用阳刚的大体量的家具和独特的陶瓷艺术与古董木片作为艺术装饰品，点缀生活的雅趣。大空间上遵循开放式设计原则，卧室空间则注重隐私为主。为了达到预期的设想，整个项目我们几乎花了两年时间才完成。

Project Name / 项目名称：Vail Getaway Home
Design Company / 设计公司：Mary Anne Smiley Interiors LLC
Designer / 设计师：Mary Anne
Project Location / 项目地点：Colorado, USA

Winter Sonata
冬季恋歌

With freedom to create without restraint and the client's trust and support, the ultimate getaway home in Vail, Colorado was created.

It was winter during the time of construction, which ended up being the inspiration. The feel and colors of winter seeped their way into the home, creating a space with elements that reflected the beauty of the soft snow and crystal ice landscape. Nature continues to flow through with accents of branches for chandeliers and table bases, along with art from local galleries in Dallas that include wall sculptures of birds and animals. The relationship among human and animals, nature, environment is perfectly interpreted in this space.

　　自由创造，是在设计此别墅时屋主给我们的设计原则。最终也在屋主的信任和支持下，我们毫无限制地完成了这个位于科罗拉多州维尔市的度假家园别墅。

　　房子的建造期在冬天，带给设计师很多灵感。于是我们把冬天的感觉和色彩渗入到别墅里面，室内的元素也以反映柔软的雪和水晶冰景的美丽为引导，整体都遵循与自然融为一体的设计原则，包括枝形吊灯、枝形摆件、以及枝形桌子底座等装饰都延续着自然风特色。除此之外，还选用了达拉斯当地画廊的艺术品，包括鸟类和动物的壁雕，将人与动物，人与自然，人与环境完美地诠释在此空间内。

The introduction of light adds glory into the space. The kitchen, a once closed-in, dark hole in the wall, was opened up to the stairwell; allowing the space to feel larger. Cabinets and countertops got a makeover with back-painted glass, making the space feel clean and open; reflecting even more light in. When the warm sunlight passes through the snow and meets the steam in the kitchen, we can see a beautiful and vivid life picture.

光线的引入和放大，给室内空间增添无穷的色彩。原有厨房是封闭的，墙上还有一个黑洞，经过休整，被扩展到楼梯间，使空间更大。橱柜和台面都被改造了，背面涂漆的玻璃使空间更干净、开放，反射出更多的光线，视线也得到进一步延伸。当窗外的暖阳，穿过层层白雪，与厨房里的热气汇合，散发出的人间烟火气息，是一幅多么美妙的画面。

The color scheme of wintry sky blue, with touches of shimmery white tile, and silvery metals were used to reflect the surrounding winter landscape. Warmth was brought into the master with a curly lamb rug, with quilted bedding. In the master bath, the shower was created with a resin material, allowing us to "bend" it into any shape. This state-of-the-art material was used in multiple rooms and for multiple uses, giving it a life of its own.

冬季的天蓝色搭配闪烁的白色瓷砖或银色金属，很好地反映出了周围的冬季景观。主卧一张卷曲的羔羊地毯，素雅的床品被褥，给这个冷冽的冬天带来了温暖感。主浴室里，树脂材料做成的淋浴，可自由转变，"折弯"变成任何形状。这种先进的材料，被以新的方式使用，用于多个房间，赋予它自己的生命。

Dancing Party in a Youthful Manse
焕染青春的庄园舞会

Design Company / 设计公司：PROjECT. interiors
Designer / 设计师：Jennifer Kranitz
Project Location / 项目地点：River Forest, IL, USA
Area / 面积：1115m²
Photographer / 摄影师：Tony Soluri

This 1924 manse, home to vibrant young newlyweds, came with a special twist: the male half of this pair grew up here and he knew its stately bones better than architect, contractor or designer. This couple's goal became clear: to preserve the integrity of the space, without getting lost in sentimentality and take every opportunity to infuse color and personality. After a new design, the old house is full of vibrancy and hope.

  这是一个 1924 年的庄园，是一对充满活力的年轻新婚夫妇的家，对于男主人来说它有一个特殊的意义：这是他长大的地方，所以他比建筑师、承包商或设计师更了解它的庄严建筑构架和历史性。这对夫妇再设计的目标是保持空间的完整性，同时又不迷失在多愁善感的状态中，准确抓住每一个机会为空间注入色彩和个性，让这座古老的庄园在新的设计情境下焕发出活力与希望！

The living room, once walls of deep stained walnut built ins and moulding, now welcomes you with white sprayed walls. Original crystal chandeliers dance on the scrolling white plaster moulded ceiling. Throughout the expansive living room, you'll spot vintage finds: hot pink pillows out of a 60s vintage dress, brass globe table lamps, vintage Hermes scarves within lucite display boxes, feather juju hat, leather and gold zipper pillows. A collection of old and new furniture create clusters for cocktails and conversation.

由于客厅原来是深染胡桃木的墙壁，看起来比较厚重，现在设计师将它以白色喷墙，可以让原始水晶吊灯在滚动的白色石膏模上尽情舞动。宽敞的客厅里，60年代的复古礼服制成的粉红色靠枕、黄铜球桌灯、透明合成树脂展示盒里的爱马仕围巾、羽毛帽、带金拉链的皮革抱枕等，新老家具的互相搭配，为屋主随时准备鸡尾酒会和闲谈提供了方便。

The exterior is surrounded by greenery, and a private pool allows the couple to enjoy fitness. The gorgeous colors of the furnishings create a fresh and leisure atmosphere. With flowers around the long party table in the outdoor garden, life will bring more surprises to the couple! External walls of the manse were finished with red bricks which reveal a historic and nostalgic feeling, and they are set off by the trees!

走出庄园，是青葱绿意的树木环绕，一旁的私人泳池，可以为健身提供很好的娱乐，绚丽的软装色彩，清新自在。露天的庭院派对长桌，鲜花环绕，将会给新婚夫妇带来更多的生活惊喜！庄园整体外墙以古老的红砖堆砌，增添了历史的厚度与怀旧的情怀。在树木的映衬下，焕然生机！

图书在版编目（CIP）数据

　　欧美别墅软装 / 深圳视界文化传播有限公司编 . -- 北京：中国林业出版社，2017.8
　　ISBN 978-7-5038-9250-9

　　Ⅰ．①欧… Ⅱ．①深… Ⅲ．①别墅－室内装饰设计－作品集－欧洲②别墅－室内装饰设计－作品集－美国 Ⅳ．① TU241.1

　　中国版本图书馆CIP数据核字（2017）第 207224 号

编委会成员名单
策划制作：深圳视界文化传播有限公司（www.dvip-sz.com）
总 策 划：万绍东
编　　辑：杨珍琼
装帧设计：黄爱莹
联系电话：0755-82834960

中国林业出版社 · 建筑分社
策　　划：纪　亮
责任编辑：纪　亮　王思源

出版：中国林业出版社
（100009 北京西城区德内大街刘海胡同 7 号）
http://lycb.forestry.gov.cn/
电话：（010）8314 3518
发行：中国林业出版社
印刷：深圳市雅仕达印务有限公司
版次：2017 年 9 月第 1 版
印次：2017 年 9 月第 1 次
开本：235mm×335mm，1/16
印张：20
字数：300 千字
定价：428.00 元（USD 86.00）